GENETICS

HISTORY OF SCIENCE

GENETICS

FROM MENDEL
TO GENE SPLICING

BY CAROLINE
ARNOLD

FRANKLIN WATTS
NEW YORK • LONDON • TORONTO
SYDNEY 1986

Diagrams by Vantage Art
Diagrams on pp. 30 and 60 from
Principles of Genetics, 3rd ed.,
Eldon J. Gardner, © 1968 by
John Wiley and Sons, Inc.
Reprinted by permission of
John Wiley & Sons, Inc.

Photographs courtesy of:
The Bettmann Archive: pp. 9, 21;
Ron Austing/Cincinnati Zoo: pp. 13, 14;
Gail Stetten, Ph.D./John Hopkins
University School of Medicine: p. 29;
Camera Press Ltd.: p. 35;
UPI/Bettmann Newsphotos: p. 41;
National Research Council: p. 44;
The National Hemophilia Foundation:
p. 51; Cold Spring Harbor Laboratory/
Research Library Archives: p. 59;
Plant Technology Corporation: p. 63;
Applied Biosystems Inc./Foster City, CA: p. 66

Library of Congress Cataloging
in Publication Data

Arnold, Caroline.
Genetics: from Mendel to gene splicing.

(History of Science)
Bibliography: p.
Includes index.
Summary: Discusses genetics discoveries and re-
search, with explanations of heredity, the contri-
butions of Gregor Mendel, DNA, and recombinant DNA.
1. Genetics—History—Juvenile literature.
2. Genetics—Juvenile literature. [1. Genetics]
I. Title II. Series: History of science
(Franklin Watts, inc.)
QH437.5.A76 1986   575.1   86-5668
ISBN 0-531-10223-8

CONTENTS

CHAPTER

WHAT IS HEREDITY?

Have you ever been told that you have your mother's eyes, your father's height, your grandfather's red hair, or been compared in a similar way to one of your relatives? Of course, you cannot really have your mother's eyes or your grandfather's red hair, but you may look as if you do. Comments such as these indicate the universal recognition that parents transmit some of their characteristics to their offspring, and that their offspring, in turn, will pass on some of these same characteristics to *their* offspring.

The study of how living organisms pass on their special features to the next generation is the science of genetics. The "father" of modern genetics is Johann Gregor Mendel, an Austrian monk, whose pioneering work with garden peas gave him the information he needed to formulate the basic rules of heredity. In February and March 1865, when Mendel first reported the results of his experiments to the Natural History Society in Brunn, Austria, the full implications were barely understood. Like many scientific discoveries, Mendel's ideas were slightly ahead of their time. It was not until the turn of the century, when other scientists rediscovered and confirmed Mendel's principles, that the science of genetics was fully born. Since then, pieces to the genetic puzzle have been contributed by many scientists including Francis H. Crick and James D. Watson, who in 1953 first described the chemical structure for heredity—DNA. Genetic research has since grown at an astounding rate and is currently one of the most exciting frontiers of science.

BEFORE MENDEL

The discoveries of Mendel were the product of both the man and the intellectual climate in which he worked. Without the work of scientists who went before him, Mendel could not have formulated his experiments nor reached the same conclusions. The eighteenth and early nineteenth centuries had seen a tremendous explosion of scientific information. It was a period of both great expansion and specialization. European explorers were traveling to far corners of the earth and bringing back with them all kinds of new plants and animals. These newly discovered creatures raised two important questions among scientists. They were, How can we distinguish one kind of animal or plant from the next? and secondly, What causes these differences?

The differences between animals such as giraffes and tigers are obvious. But, it is not so obvious whether animals such as fox terriers and sheep dogs are different kinds of animals or just two varieties of the same kind. The person who emerged to solve this problem was a Swedish scientist, Carl Linnaeus.

Originally trained as a physician, but far more interested in botany, Carl Linnaeus became, in 1732, the plant collector of the Uppsala Academy in Sweden. His first collection trip took him across Lapland, all the way to the Arctic Circle, and on this trip he collected hundreds of previously unknown plant specimens. After he returned, he named and classified these plants within a system that grouped the plants according to the shape of their reproductive parts. Later, as he traveled around Europe pre-

Carl Linnaeus, in Lapland dress

senting his findings, scientists were impressed with his new system. Although some criticized his method as "artificial," it was nonetheless the best way yet devised to put the otherwise confusing array of living things into some kind of order.

The system of naming that Linnaeus used, and which we still use today, gives each living thing two names, both in Latin. The first name, always capitalized, designates the name of the genus to which the animal belongs. The second name designates the species. The scientific name for modern man, for instance, is *Homo sapiens,* "*Homo*" being the genus and "*sapiens*" the species. Linnaeus grouped similar genera (the plural of "genus") into orders, and orders into classes. Man, for instance, belongs to the order of mammals and the class of vertebrates (animals with bony skeletons). Additional subdivisions, such as phylum and family, have been added since the time of Linnaeus, but the structure of his system remains essentially the same today. In his lifetime Linnaeus attempted to identify and classify every living thing, a gargantuan task which filled ten volumes of his book *Systema Naturae.*

At the time Linnaeus worked it was popularly believed that the number of different kinds of creatures on earth had been fixed since the time of biblical creation and that no animal had become extinct since the biblical Flood. Growing evidence, including Linnaeus's own exhaustive catalogue, suggested that this could not be true. The discovery of extinct plants and animals in fossils proved that some organisms had indeed disappeared since the Flood. Also, an entirely new species of plant would occasionally appear in the midst of a farmer's field. These new plants, called "sports," seemed to emerge spontaneously. No one could explain how they got there, but it was clear that creation was an ongoing process.

By the end of the eighteenth century a number of scientists, including Erasmus Darwin, the grandfather of Charles Darwin, had become convinced that the number and form of species was not fixed. However, no one had yet come up with an explanation of how or why these changes occurred.

One scientist who was interested in how species changed over time was Jean Baptiste de Monet Lamarck, a French botanist and zoologist. Lamarck was the first person to conceive of the unified study of plants and animals, and as such is the originator of our modern idea of biology.

Like many scientists of his time, Lamarck viewed the world of plants and animals as a kind of ladder on which less developed creatures occupied lower rungs and more highly developed creatures sat on the top. However, Lamarck's departure from the traditional view was that he claimed that each species "strived" to improve itself and move up the ladder. He suggested that they did this by adapting to their environment and then passing these characteristics on to the next generation.

One of his examples is how the giraffe got such a long neck. Lamarck believed the long neck evolved over many generations in which giraffes stretched their necks further and further to reach the leaves on ever higher branches. Mating giraffes would produce babies with necks slightly longer than theirs. Although Lamarck's ideas are generally ridiculed today, their influence was felt well into the twentieth century. When one realizes that in Lamarck's time no one understood either the mechanism of reproduction or the transmission of characteristics through genes, it is easy to see how his ideas could be accepted as reasonable.

HEREDITY AND REPRODUCTION

A knowledge of heredity is closely tied to an understanding of the reproductive process. Perhaps the most eminent early scientist with views on reproduction and heredity was Aristotle. Like other Greek philosophers and scientists, he believed that each offspring was a mixture of elements from each parent's body. Aristotle's theories and those like it are known as the *particulate theories* of reproduction. They state that particles of the parents' bodies somehow mix and rejoin to form the new organism. Aristotle claimed that in animals blood was the key ingredient that was passed from one generation to the next. His beliefs dominated scientific thinking for hundreds of years and form the roots of such expressions as "pure blood," "blood relative," or "bloodline," which are still used today.

Basic to the particulate theorists is the idea that each organism is a blend of its parents' characteristics. In ancient times such a theory was used to explain the existence of some of the more unusual animals on earth. The giraffe, for instance, was supposed to have resulted from the mating of a camel and a leopard, and the ostrich from the combination of a camel and a sparrow. Although a cross between different kinds of animals is possible, it is the rare exception rather than the rule. Also, such crosses are possible only between *related* species, such as between a horse and a donkey to produce a mule, or a hinny, or between a lion and a tiger to produce a "liger" or a "tigon."

During the Renaissance, experiments by the English physician William Harvey (1578–1657) demonstrated the development of an organism from an egg inside its mother's body. Clearly there was no mixture of blood, contrary to what Aristotle had stated. However, Harvey mistakenly gave all credit for the reproductive process to the mother, or the female of the species. He suggested that the egg contained all the material necessary to enable the new organism to grow.

A variation of this view was popular in the eighteenth century. It promoted the idea that the entire organism was contained not in the egg, but in the sperm, provided by the male. The development of the microscope in the 1680s by the Dutch scientist Anton van Leeuwenhoek had made it possible for people to view tiny sperm for the first time. Some imaginative scientists began to claim that they could see miniature bodies curled up inside the head of the sperm. These little bodies were then supposed to grow into embryos inside the womb. The idea that an organism's offspring preexisted in miniature in the egg or sperm of its parent was called the *preformation theory*.

The baby bongo shown here with its surrogate
mother, an eland, is the result of an inter-
species, cross-genera embryo transfer.

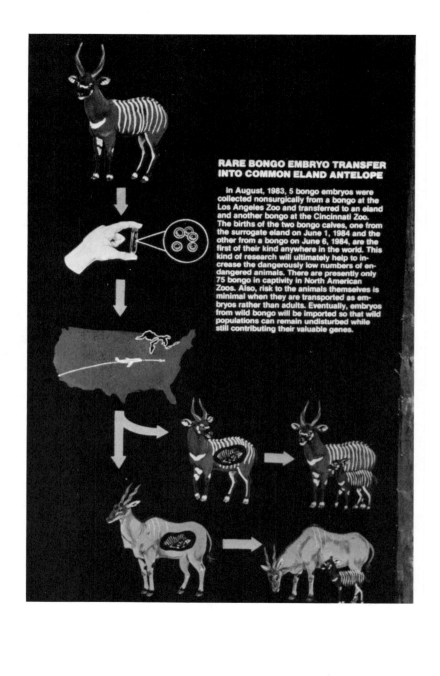

RARE BONGO EMBRYO TRANSFER INTO COMMON ELAND ANTELOPE

In August, 1983, 5 bongo embryos were collected nonsurgically from a bongo at the Los Angeles Zoo and transferred to an eland and another bongo at the Cincinnati Zoo. The births of the two bongo calves, one from the surrogate eland on June 1, 1984 and the other from a bongo on June 6, 1984, are the first of their kind anywhere in the world. This kind of research will ultimately help to increase the dangerously low numbers of endangered animals. There are presently only 75 bongo in captivity in North American Zoos. Also, risk to the animals themselves is minimal when they are transported as embryos rather than adults. Eventually, embryos from wild bongo will be imported so that wild populations can remain undisturbed while still contributing their valuable genes.

Today we know that neither the particulate nor the preformation theorists were right, but rather that both parents have specialized body cells designed for reproduction, and that these cells contribute equally to the development of offspring.

By the middle of the nineteenth century, scientific observation through the microscope had established the cell to be the basic building block of life. Single-celled organisms such as amoebas and bacteria had been observed to multiply simply by dividing themselves. This is called *asexual reproduction* and requires only one parent cell, which produces offspring identical to itself. Multiplying yeast is an example of this kind of reproduction.

Most plants and animals, including man, need two parents—one male and one female—to reproduce, each of them providing one half of the reproductive cell. This is called *sexual reproduction.* The female of each species produces egg cells, and the male produces sperm cells. In plants, the male cells are contained in grains of pollen. (Some plants have both male and female cells on the same plant.) Before joining, the male and female reproductive cells split in half. Each of the split cells (called *gametes*) contains half the parent's hereditary material.

When a male gamete unites with a female gamete, the egg becomes fertilized. After fertilization the cell divides and multiplies as in asexual reproduction. As the organism develops, cells begin to specialize to form body parts with different functions and different characteristics.

It is the potential for new combinations of characteristics in the new organism as a result of sexual reproduction that creates the possiblity of genetic diversity. When a female and male mate and produce young, the offspring will have a new combination of their parents' characteristics. Each individual will be unique. When they reproduce, their offspring will have yet a different combination of characteristics. When a new combination of characteristics helps an organism to better adapt to its environment, it is more likely that those characteristics will be passed on to future generations.

GREGOR MENDEL

Today, a knowledge of the principles of heredity is so basic to our fundamental understanding of all of the biological sciences that it is hard to believe that it has been little more than one hundred years since these ideas were first formulated. It is perhaps more amazing that their discovery was made not by an eminent scientist doing experiments at a prestigious university, but rather by a humble monk working in the monastery vegetable graden. That monk's name was Johann Gregor Mendel.

MENDEL'S EARLY LIFE

Mendel was born in 1822 in the Moravian village of Heinzendorf, now part of Czechoslovakia but then part of Austria. His parents, Anton and Rosina Mendel, named him Johann. Later, in 1843, when he entered the monastery and was required to choose a new name, he became known as Brother Gregor Mendel.

From his father, a poor but hardworking farmer, young Johann learned farming and a love for horticulture, or plant breeding. Under ordinary circumstances he would probably have grown up to follow in his father's footsteps, as did most other farmer's sons in those days. However, Johann was an avid student and did well at the village school. There he was influenced by a teacher named Thomas Makitta, who taught a class on the techniques of fruit culture and hybridization. During the late eighteenth and early nineteenth centuries there was great interest in the production of hybrids, that is, of growing plants and animals with special characteristics by mating two purebred species. Although the details of how hybrids developed was not known, the advantages of hybrids were clear. For instance, hybrid wheat strains, which were resistant to disease, could be used to produce significantly larger yields. Many people became interested in breeding hybrids as a hobby, and Mendel was one of them.

After finishing the village school, Mendel went on to high school in the neighboring village of Troppau. Unfortunately, Johann's family was too poor to be able to properly support him there, and Johann became ill due to improper nutrition and was forced to drop out of school. Then, just when he had regained his health and was ready to return to his studies, his father was badly injured by a falling tree.

Anton Mendel's injuries made it impossible for him to continue working the family farm. It appeared that this would be the end to Johann's education. However, after the farm was sold and Johann's father distributed the proceeds to Johann and his sisters, Johann discovered that, with the help of one of his sisters, he would be able to go to school after all. So, for the next four years he attended the Philosophical Institute at Olmutz and received a degree in philosophy. After this he still wanted to continue his learning, but by this time had run out of money. He consulted one of his teachers at the Institute, Professor Michael Franz, and on his advice decided to enter the Augustinian monastery in the town of Brunn. There he would be able to live a useful life and could continue his studies on his own.

When Mendel arrived at Brunn, he admired the extensive and lovingly tended gardens which surrounded the monastery. These were the work of an older monk, Father Aurelius Thaler. When Father Thaler died shortly after Mendel arrived, Mendel became the caretaker of these gardens, and he soon settled down to a pleasant life of monastic study and plant tending. In addition to his work at the monastery Mendel also took a part-time job as a substitute teacher of natural science at a nearby high school. After several years of teaching he decided to take the examination for certification as a full-time teacher. He failed miserably.

Most of Mendel's scientific knowledge up to that point was based on his own experience, for his formal education had involved mostly classical studies. To correct this gap the monastery decided to send Mendel to the University of Vienna, then a major center of scientific studies. There Mendel could take courses in zoology, botany, microscopy, physics, and chemistry. One of his teachers at the University was Franz Unger, who encouraged Mendel's natural curiosity and questioning of traditional views. Later, Professor Unger would become a champion of Charles Darwin's views on evolution.

When Mendel felt himself sufficiently prepared, he again took the teacher's examination, and again he failed. One of his examiners wrote that his problem was that he did too much original thinking! Some people feel that Mendel may have been discriminated against because he was a Catholic priest. Whatever the reason for his failure, this was a turning point in Mendel's life.

Mendel returned to the monastery and continued his part-time substitute teaching. There, encouraged by the abbot, he resumed his study of plant breeding in a small garden plot near the main monastery building. The experiments he did there would eventually change the course of scientific history.

THE EXPERIMENTS
IN THE GARDEN

The question which most intrigued Mendel was, How can we explain the multitude of different shapes and colors of all living things? Mendel's previous experience with plant breeding, both on his father's farm and at the monastery, had acquainted him with the enormous variations that can occur in a single kind of plant. Mendel began planning his experiments, and started by reading books and scientific journals to find out what other people had already discovered in this area. Mendel found that the results of most previous studies had been confusing and inconclusive. However, he noticed that none had been done for an extensive period of time, that most people had tried to draw their conclusions solely from the appearance of the plant or animal, and that no one seemed to have tabulated the results. What would set Mendel's experiments apart from those earlier studies was the patience and thoroughness with which they were done, their focus on just a few variable characteristics, and their foundation in mathematics.

At the monastery, Mendel was free from outside pressures to finish his experiments quickly. This gave him the time to do his work carefully and completely. After some initial experiments with white and gray mice, which had given unsatisfactory results, Mendel decided to use the common garden pea for his project. The advantages of the pea included the fact that numerous hybrid varieties already existed and could be obtained easily from seed companies. Mendel initially ordered thirty-four varieties of seeds in order to select from them those which would be used in his experiments. Then he planted these seeds for two successive years to make sure that they were truly purebred.

Mendel was not the first scientist to experiment with garden peas. Thomas Knight of the British Horticultural Society, working nearly a century earlier, had written about the virtues of the pea as an experimental plant. He wrote that he used it ''not only because I could obtain many varieties of the plant of different forms, size, and colors, but also because the structure of the blossoms, by preventing the ingress of insects and adventitious fauna, has rendered its varieties remarkably permanent.''

An important feature of the pea plant is that the structure of the flower makes it easy to control fertilization. Like most plants, the pea plant produces seeds from which new generations can grow. Each seed is formed by the union of eggs and pollen in the flower of the plant. This union is called *fertilization.*

In some plants the eggs and pollen are on different flowers or even on different plants. However, on the pea plant both the pollen and eggs are in the same flower and the plants are *self-fertilizing*. The tiny grains of yellow pollen from the tip of the stamen find their way down the tubelike pistils, each having an egg at its base. As the pea plant grows, each of these grows into a pea seed inside the pea pod. Because all the eggs and pollen from each plant are the same, all the peas in each plant are identical. Unlike the flowers of many other kinds of plants, the pea flower remains closed during the fertilization period, thus preventing pollen from other plants fertilizing the eggs.

In order to *crossbreed* two different kinds of peas, in other words to fertilize the egg of one plant with the pollen of another, Mendel and his helpers had to painstakingly open the flower of each plant by hand. Then on some of the flowers they carefully snipped off the stamens so that pollen could not form. From another variety of plants Mendel then collected pollen to fertilize the first group. Then to prevent any chance fertilization by pollen from other flowers, they tied tiny calico bags over each flower. It was a tedious job, and only the beginning of a long experimental process. During the eight years that Mendel conducted his experiments, he worked with and recorded the results from ten thousand different plants!

One of the secrets of Mendel's success is that he chose to focus on only one or two variables at a time. He selected seven characteristics to study.

1. The color of the seed: whether it was yellow or green.
2. The shape of the seed: whether it was round or wrinkled.

Johann Gregor Mendel experimenting with sweet peas in the garden of an Augustinian monastery in Brunn, Germany

3. The color of the flowers: whether they were white or red.
4. The form of the ripe pod: whether it was smooth or wrinkled between the seeds.
5. The color of the unripe pods: whether they were green or yellow.
6. The position of the flowers: whether they were distributed along the stem or bunched at the top of the stem.
7. The length of the stem: whether it was long or short.

For his first experiment Mendel crossbred plants with round and wrinkled seeds. In one part of the garden he planted round seeds; in a separate area he planted wrinkled seeds. When the round-seed plants grew and developed flowers, he snipped off the stamens and sprinkled the flowers with pollen from the stamens of the wrinkled-seed plants. On the flowers of the wrinkled-seed plants he sprinkled pollen from the round-seed plants. He did this on a total of 287 flowers on seventy plants. Then he waited for the plants to grow.

Many weeks later, when the seeds were ripe, Mendel harvested his crop. As he and his assistants opened the pods and inspected the seeds, they found, much to their amazement, only round seeds. Not a single wrinkled seed appeared!

Mendel's results were contrary to all the current ideas about inheritance which asserted that the characteristics of each parent would blend in the offspring. If that were true, then Mendel's seeds should have been partially wrinkled, that is, halfway between round and wrinkled. Mendel's other experiments produced results similar to his first. For instance, when tall and short plants were crossed and the seeds were planted, only tall plants resulted. There were no medium-height plants. Mendel called the offspring of the first cross the *first generation* and the characteristic that all the plants shared was said to be *dominant*. The characteristic that had disappeared he called *recessive*. Round seeds and tall plants were dominant characteristics, whereas wrinkled seeds and short plants were recessive characteristics.

Mendel was curious to see what would happen when he planted the seeds of the first generation to produce a *second generation*. The next summer he planted a plot of round seeds and patiently waited for the plants to grow and mature. This time, when he opened the pods to examine the seeds, he found he had both round *and* wrinkled seeds. The recessive characteristic had reappeared in the second generation! When Mendel counted the seeds he found that the round seeds outnumbered the others by almost exactly three to one. His other experiments produced a similar ratio in the second generation.

Mendel then planted his second generation seeds to produce a *third generation*. Upon harvesting the plants, he found that all the wrinkled-seed plants had produced only wrinkled seeds. Yet, the round-seed plants had produced a mixture of round and wrinkled seeds, again in a ratio of three to one.

The following winter Mendel sat down to analyze the information he had collected. He had planted and counted thousands of peas. In each experiment he had looked at only one characteristic that appeared in two distinct forms. Since he knew that each plant was the result of pollen and egg combining to produce the seed, he concluded that the factors that determined the appearance of the plant must be in these reproductive cells. Because it was clear that each characteristic was the result of the combination of two factors, or two components, one contributed by each parent, Mendel correctly assumed that each parent's reproductive cells contained only one of these factors.

In purebred plants all the reproductive cells would have the same factors for the purebred characteristic, but in crossbred plants the reproductive cells would have some of each factor. For instance, a purebred round-seed plant would produce pollen and eggs with only a round-seed factor. Similarly, purebred wrinkled-seed plants would produce wrinkled-seed eggs and pollen. When the plants were crossbred, then one wrinkled-seed factor would combine with one round-seed factor. The new seed would be round because round is dominant.

Mendel devised a shorthand mathematical system to chart

his cross breeding. He decided to use single letters to represent each factor. A capital letter denoted the dominant factor, and the same letter lowercased denoted the recessive factor. An X meant "crossed with." Here is how he showed the first generation cross between round (R) and wrinkled (r) seeds:

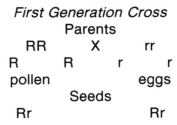

First Generation Cross
Parents
RR   X   rr
R   R   r   r
pollen   eggs
Seeds
Rr   Rr

When Mendel used his system to chart the second generation cross, his results were amazingly similar to those that he had achieved in his experiments.

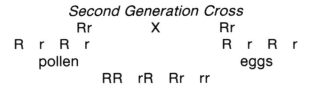

Second Generation Cross
Rr   X   Rr
R r R r   R r R r
pollen   eggs
RR   rR   Rr   rr

Because the round-seed factor (R) is dominant, it appears three-quarters of the time. The wrinkled seed appears only when both factors are recessive.

In the next year Mendel crossbred round, yellow seeds with wrinkled, green seeds. He wanted to see if one factor affected the appearance of another factor. After harvesting and counting the peas he found that he had four different kinds: round yellow, round green, wrinkled yellow, and wrinkled green. They appeared in a ratio of 9:3:3:1. When he looked at the seed coat and color separately he found that they appeared in the same three-to-one ratio that he had seen before. Clearly, in this case, the factors seemed to be independent.

When Mendel had completed all his experiments, he formulated several rules, or "laws," of heredity:

1. Heredity is determined by elements contained in the two cells contributed by the parents of the organism, and these elements combine randomly.
2. Each characteristic of an organism is determined separately from the others.
3. When two different characteristics combine, one characteristic will dominate over the other. (This theory turned out not to be generally applicable.)

Finally, on a cold winter night in February 1865, Mendel took all his information to a meeting of local intellectuals called the Brunn Society for the Study of Natural Science. Because his report was so long, Mendel could not finish in one evening and completed the second half at a meeting a month later. During both talks, the members appeared to listen politely to what Mendel had to say. Yet, when he finished, there was not a single question and it appeared that no one recognized the importance of his discoveries. Nevertheless, they invited Mendel to publish his paper in the Society's journal, which he did in 1866 under the title "Experiments in Plant Hybridization."

Despite the cool reception in Brunn, Mendel hoped that his paper, once published, would attract the attention of biologists elsewhere. He was disappointed again, for no one responded. Mendel then sent his paper to the eminent botanist Karl Wilhelm von Nägeli, but even he failed to recognize the implications of Mendel's work, perhaps because he had already developed his own views about inheritance.

In 1868 Mendel was elected to be the director of his monastery, and administrative duties prevented further experiments with plants. He became active in the community, and, when he died in 1884, the whole city of Brunn mourned the loss of their kind abbot. No one suspected that one day he would be looked upon as the founder of modern genetic science.

CHAPTER

MENDEL
REDISCOVERED

When the English scientist Robert Hooke looked at a slice of cork under his microscope one day in 1665, he noticed that it was divided into small regular chambers. Because the pattern resembled an overview of a monastery plan, in which there are many small rooms called *cells,* he named these cork units cells, too. As other scientists looked at other plants and animals under the microscope, they found out that all living things are composed of cells. The *cell theory* of life was finally developed in 1838 by two German scientists, Matthias Schleiden and Theodor Schwann.

The basic unit of all life is the cell. Our bodies are made up of billions of different kinds of cells, each one so small that you cannot see it without a microscope. Each living cell is filled with a jellylike substance called *protoplasm*, and in the center of each cell is a round structure called the *nucleus.* As the knowledge of the workings of the cell expanded in the nineteenth century, it provided the basis for understanding the mechanism of heredity.

When Mendel conducted his experiments with peas, he knew that the hereditary substance, that is, the directions for the development of the new plant, must be contained in the reproductive cells of the parent plant, since those were the cells that produced the new seeds. Yet, since this hereditary substance was not visible, Mendel was able only to assume its existence. Then toward the end of the nineteenth century, scientists discovered that there were even smaller units inside the cell nucleus. These were tiny, thin threadlike structures. They named them *chromosomes* because they were easily stained with colorful chemical dyes. (*Chromo* means "color" in Greek.)

One thing that intrigued scientists about chromosomes was their behavior during cell division. Most cells have the ability to reproduce themselves by dividing. Just before division, the cell takes in foodstuffs and converts them to substances and structures normally contained in the cell, making the cell double in size. All the chromosomes duplicate and arrange themselves along the line of division. Then, when the cell divides, each new cell, called a *daughter cell*, gets one duplicated half of each chromosome. The chromosomes of the daughter cells are exactly like those of the parent cell in size, shape, and number.

The discovery of chromosomes was just one of several important pieces in solving the larger puzzle of heredity. As late as 1870 the basic process of reproduction was still being hotly debated. Many, including Charles Darwin, maintained that cells from an entire organism influenced the reproductive cells. At this time a German scientist named August Weismann was studying simple sea animals, including jellyfish and hydras. He noticed that the reproductive cells of these animals were distinct from those of the rest of the body. Weismann was interested in how organisms inherited characteristics from their parents and in 1892 published a paper in which he described his "theory of the continuity of germ plasm." He believed that all the information for the new organism was contained in the reproductive cells contributed by the parents. As he elaborated his theory in following years, he asserted that this "hereditary substance" was contained in the chromosomes. Weismann's ideas were contrary to most current scientific thought and stimulated controversy. Although many of his ideas later proved to be erroneous, his idea that the single cell contains all the necessary information for the next generation was a contribution that paved the way toward the rediscovery of Mendel's work.

One scientist who was influenced by Weismann's work was a Dutch botanist, Hugo de Vries. De Vries was doing a series of plant-breeding experiments. Like Mendel, he crossed two varieties of plants, in this case of the silene plant, and recorded the proportion of plants in which certain characteristics appeared in second and third generations. In 1900, after seven years of work, he was ready to publish his report. Only then did he discover that Mendel had made the very same discoveries nearly forty years earlier.

At the same time two other scientists, Carl Correns in Germany and Erich Tschermak von Seysenegg in Austria, were also conducting plant-breeding experiments. They, too, got similar results to Mendel's from crossing hybrids. All three of these men became aware of Mendel only when they researched the literature in preparation for their own reports. When they published their findings and referred to Mendel's research, Mendel's discoveries began to be appreciated at last.

Although de Vries, Correns, and Seysenegg are usually given the credit for "rediscovering" Mendel, it was another man, William Bateson, an English scientist, who was responsible for promoting Mendel's work. Bateson lectured widely and also conducted his own research to find support for Mendel's ideas. Bateson suggested a new name for this rapidly emerging scientific field: that name was "genetics."

A Danish biologist, Wilhelm Johannsen, who worked with inheritance in beans, was the first to use the word "gene" to describe the hereditary substance. Johannsen also coined the words *phenotype*, meaning the appearance of an organism, and *genotype*, meaning the genetic factors present in an organism.

In 1902, W. S. Sutton, an American biologist working at Columbia University in New York, recognized the relationship between Mendel's work and chromosome behavior. Sutton formulated the *chromosome theory,* which maintained that the he-

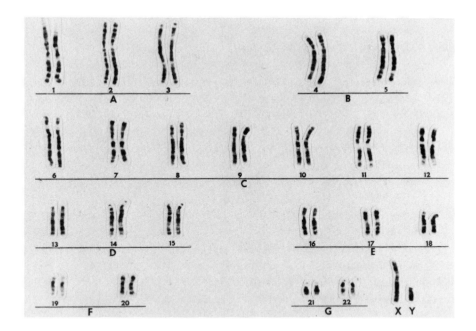

Paired chromosomes from the cell of a human male. The X and Y chromosomes at the lower right determine the sex.

reditary substances, or genes, were on the chromosomes or were parts of the chromosomes. This explained why genes occur in pairs (because they are on matching pairs of chromosomes), why each member of the pair is derived from one parent (because one of each chromosome pair is contributed by a parent), and why the genes separate during cell division (because the chromosomes separate).

One of Sutton's teachers at Columbia, C. E. McClung, was studying the chromosomes of a small insect called *Pyrrhocoris*. One of the chromosomes of this insect behaved so strangely that McClung named it the X chromosome. Because the X chromosome seemed to have no matching chromosome in males, it was

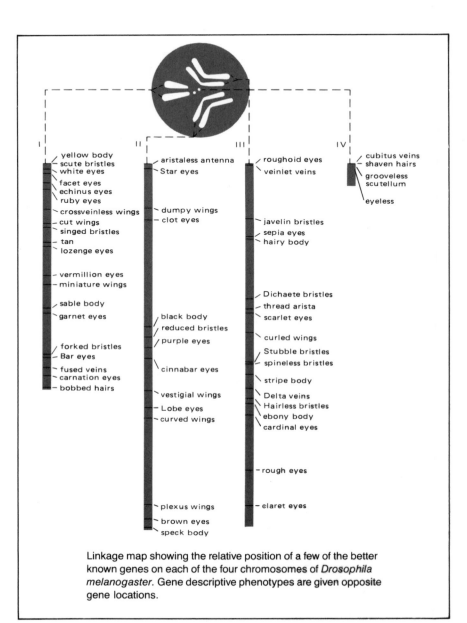

Linkage map showing the relative position of a few of the better known genes on each of the four chromosomes of *Drosophila melanogaster*. Gene descriptive phenotypes are given opposite gene locations.

thought at first that this was the chromosome that determined the sex of the insect. It was not until 1910 that the tiny Y chromosome was discovered and found to be the sex determining chromosome. In mammals males are determined by one X and one Y chromosome, and females by two X chromosomes. At this time the focus of genetic research turned to the chromosome and many scientists began to try to determine the exact structure of the genes on them.

THE FRUIT FLY

Two men whose work laid the foundation for much of the genetics research during the early twentieth century were Thomas Hunt Morgan at Columbia University and W. E. Castle at Harvard University. Morgan's group at Columbia included scientists Edmund Beecher Wilson, Calvin Bridges, A. H. Sturtevant, and H. J. Muller. Working between 1909 and the mid-1930s, all of these scientists made many basic discoveries concerning how genes work.

Both Morgan's and Castle's groups studied a tiny insect called *Drosophila melanogaster*, commonly known as the fruit fly. This organism is uniquely suited to genetic research for a number of reasons. First, it is small and takes up little laboratory space; second, it reproduces rapidly and in large numbers; and third, it requires very little care. Best of all, each fly has only eight chromosomes, all readily visible under the microscope.

It was obvious that if each characteristic of an organism was controlled by a gene, and there were only eight chromosomes in a fruit fly, then each chromosome would have to contain many genes. Because certain genes always came together on the same chromosome, these genes were always passed on together to the next generation. This concept came to be known as *chromosome linkage* (see illustration on page 30).

In 1910, Morgan noticed that one of the flies in the bottle had white eyes instead of the usual red eyes. This white-eyed fly was a male. When the white-eyed male was mated with a red-eyed female, all the offspring had red eyes, indicating that the red eyes were a dominant characteristic. When the second generation was bred, the red-eyed flies outnumbered the white-eyed flies in a ratio of three to one, as expected. However, the unusual result was that *all* white-eyed flies were male. Thus, Morgan reasoned, the gene for eye color must be on the X, or sex, chromosome. This was the first time anyone had placed a specific gene on a specific chromosome. The term *sex linkage* is used to describe a genetic characteristic associated with a sex chromosome. Morgan later also discovered genes for yellow body color and miniature wings on the X chromosome in fruit flies.

One of the problems of genetic researchers both before and after Mendel was the confusing combinations of characteristics in the second and third generations. Characteristics were often linked, but not always. Morgan then proposed a theory to explain how this might happen. Sometimes during cell division the dividing chromosomes overlap or wrap around each other. If they break and join to the other chromosome crossing over, then there is a new arrangement of genes. Sturtevant suggested that genes are more likely to cross over when they are further apart on the chromosome, and that this information could be used to locate the arrangement of genes on the chromosome. This idea became the basis for "genetic linkage studies"—that is, locating the relative position of genes on the chromosomes.

As the people in Morgan's laboratory continued their work with fruit flies, they developed a theory about how chromosomes functioned, which they published in 1915 in a book called *The Mechanism of Mendelian Heredity.* Much of genetics research in the first half of the twentieth century focused on identifying and locating specific genes on the chromosomes.

CHAPTER

CHAPTER

THE GENETIC CODE

Although experiments with fruit flies and other organisms during the early part of the century provided important information about the effects of genes and their location on the chromosomes, the chemical nature of the gene was not yet understood. Work toward solving this puzzle began in 1928, when Frederick Griffith, a British scientist, was studying pneumonia bacteria in mice. He found two types of bacteria cells—one surrounded by a smooth coating, and the other with a rough coat. The rough-coated bacteria were mutants of the more common smooth-coated cells.

In further work on the pneumonia bacteria, Oswald T. Avery and his colleagues at the Rockefeller Institute for Medical Research in New York, analyzed the genes of the bacteria chemically and found that they were made of deoxyribonucleic acid (DNA). This was then confirmed in 1952 by Alfred Hershey and Martha Chase at the Cold Spring Harbor Laboratory on Long Island. It was already known that DNA (or a closely related compound, ribonucleic acid, or RNA), exists in all living cells, including reproductive cells, and is passed from cell to cell when they divide. Now it was declared that DNA was the chemical transmitter of genetic information.

DNA is one of the largest known molecules. One molecule of human DNA, for instance, weighs about a hundred thousand times more than a molecule of sugar. Each strand of DNA is less than a trillionth of an inch thick. Normally, the DNA is tightly folded into a tiny knot inside the cell, but if it were unraveled, one molecule would stretch to over 6 feet (1.8 m) in length!

WATSON AND CRICK AND
THE STRUCTURE OF DNA

The discovery of the structure of DNA, made by James S. Watson and Francis Crick in 1953, is perhaps the most significant biological breakthrough of this century. For their work they received the Nobel Prize in Physiology and Medicine in 1962, along with biophysicist Maurice H. Wilkins, who had worked with Crick at Cambridge University. At the time, the Royal Caroline Institute of Sweden, which awards the Nobel Prize, noted that the discovery had "no immediate practical application, but determining the molecular structure of the substance that is responsible for the forms that life takes is a discovery of tremendous importance." Little did they know that within two decades the possibilities for practical applications of this information would seem limitless.

The structure of DNA is only visible through microscopes which use X rays rather than light to illuminate the object. This technique, known as X-ray crystallography, was first used in 1912 to look at salt. By the early 1950s, British scientists Rosalind Franklin and Maurice Wilkins, had each taken high-resolution pictures of DNA. However, even these pictures did not absolutely identify the components of the DNA.

Watson and Crick used mathematics and their knowledge of biochemistry to build a theoretical model which matched the image in the X-ray microscope. Their model, called a double helix because of its double-spiral shape, looked like a twisted ladder. The building blocks of the ladder were four chemicals called bases. Each rung of the ladder was formed by a pair of the bases joined together.

James Watson and Francis Crick with their model of the structure of DNA. They received the Nobel Prize in 1953 for their discovery.

The four bases in DNA are chemicals called *adenine, guanine, cytosine,* and *thymine* (abbreviated A, G, C, and T), and each strand contains up to twelve billion of these bases. The double strands of the helix join because the bases form partners. Adenine always pairs with thymine, and cytosine always pairs with guanine. Thus, each strand of the helix forms a mirror image of the other. For instance, if one side of the helix has bases in the order AGCCT, then the other side will be TCGGA.

Each of the bases on the DNA acts like a letter of the alphabet, and just as we use letters to form words, so the bases can be used to code genetic messages. Even though there are only four bases in DNA, they can be arranged in many different orders. Along a strand of DNA, the bases are "read" in groups of three, each group being called a "codon." There are sixty-four possible codons, and these are used to direct the arrangement of small molecules called *amino acids.* Amino acids are the building blocks of proteins.

Watson and Crick's model of the DNA molecule provided the answer to how chromosomes duplicated themselves during cell division. Although the chemical bonds between the bases along each strand of DNA are strong, those across the helix are weak. In preparation for cell division, these bonds—the rungs of the ladder—break. It is like a zipper gradually coming unzipped. At the same time, within the cell, there are loose single-base molecules. In the same way that the bases of the two strands had been attracted to each other—A to T, and C to G—these loose bases attach themselves to sites along the unzipped DNA, forming two new zippers. When the split is complete, there are two identical DNA helices (the plural of "helix"), one designated for each daughter cell.

USING DNA TO BUILD PROTEINS

Inside the cell, the DNA acts something like an instruction manual, providing all the information needed for the cell to function.

However, the actual work of creating the proteins the cell needs is done by another molecule, ribonucleic acid, or RNA. RNA is like DNA in that it is a chain of bases, although its chain is shorter and often has only one strand instead of two. RNA is formed when an enzyme directs part of the DNA on the chromosome to be copied. The RNA then selects amino acids from within the cell and attaches them to each other to form the protein.

At first, it was believed that the RNA code exactly matched that of the DNA from which it was made. However, several scientists working with RNA got puzzling results when they tried to match it with the DNA from which it had been formed. In many cases the RNA was significantly shorter—as if part of the message had been deleted. By 1977, several groups of researchers realized that part of the genetic message in the DNA seems to be extraneous and is edited out before the RNA manufactures the protein. As yet, the use of the deleted portions is not known.

GENETIC MISTAKES

Just as even the most efficient typist can insert errors into a manuscript by occasionally adding or omitting a letter in a word, so can the genetic code be changed by altering the sequence of bases in the gene. Since the code is read in groups of three, the addition or subtraction of just a single base can alter the reading of each group after the mistake.

When a gene becomes changed through base changes it is called a *mutation.* When a mutation occurs in an egg or sperm cell, the new gene is passed on to succeeding generations.

Sometimes changes in the base sequence occur spontaneously when the DNA is being copied. This is rare, occurring only about once for every ten million bases copied. However, under certain circumstances, the rate of mutation can greatly increase, such as when the body is exposed to chemicals like mustard gas or to some kinds of radiation (X rays, ultraviolet light, and gamma radiation).

Usually a mutation deactivates a gene. However, since most plants and animals have two sets of genes, a mutation of one may not be noticed. However, if both genes have mutated, then the enzyme which that gene makes will be absent or malfunctioning. In some cases the consequences of a missing enzyme are relatively harmless, causing, for instance, different-colored petals on a flower or different-shaped ears on an animal. In other cases, such as in the genetic disease galactosemia, the inability to digest milk sugar, the absence or malfunction will have a serious effect.

The body's defenses against harmful mutations are special enzymes which "check" the accuracy of the base sequences of the DNA before it reproduces. These enzymes work a bit like cellular proofreaders. If one of them finds a mistake, it can replace that portion with the correct bases. However, even with this "proofreading" system, a few mistakes still slip through. Also, sometimes a mistake will be reproduced if the checking enzyme itself has mutated.

GENE SWITCHING

At the same time that Watson and Crick's discovery of DNA answered questions about the structure of the genes, it created many new questions about how this system worked. One of the major puzzles was how the genes were controlled within the organism.

In the human body, for instance, each cell contains the same set of about fifty thousand genes. Yet different genes are active in different cells. Scientists wanted to know why, for instance, do the genes in heart cells make proteins for muscle contraction but not for the making of insulin or adrenalin? The heart contains genes to make insulin or adrenalin, but these genes appear to be "turned off."

Two French scientists studying this problem, Francois Jacob and Jacques Monod, theorized that in every cell there were substances that were able to switch certain genes off. Such substances are called *repressors.*

Jacob and Monod's work was done on an organism called Escherichia coli (usually called E. coli), a bacteria which normally lives inside the intestine, but which can also be grown easily and cheaply in a laboratory in tubes or dishes filled with nutritious substances. Another advantage of E. coli is that it reproduces very fast—a new generation grows every twenty minutes.

Jacob and Monod found that when the E. coli was placed in a solution containing lactose, a sugar found in milk, it began to make large quantities of an enzyme that would convert the lactose to another sugar, glucose, which the bacteria uses for energy. However, when the E. coli was in a solution that did not contain lactose, the enzyme was not produced. They concluded that most DNA messages are turned off most of the time. Monod and Jacob and another scientist, Andre Lwoff, shared the Nobel Prize for Physiology and Medicine in 1965. Lwoff had studied the way certain viruses can live inside E. coli.

In 1965, a young scientist at Harvard University named Mark Ptashne began to study genetic repressors. So did another Harvard scientist, Walter Gilbert, and at about the same time both isolated these genetic repressors and showed them to be proteins. As they studied repressors further, they discovered that they do not just turn off genes; a repressor can also turn on the manufacture of its own gene. There are also repressors which turn off repressors. Repressors allow a cell a means of controlling the rate at which proteins are made, and thus enable them to produce a given protein for exactly as long as it is needed.

BARBARA McCLINTOCK
AND "JUMPING GENES"

Until the late 1970s it was believed that each gene had its own place on the chromosome, and that this position did not change from one generation to the next. Now scientists know that genes can spontaneously move from one site to another on the same chromosome, or even to another chromosome.

Pioneering work done by Barbara McClintock of the Carnegie Institution of Washington's Department of Genetics at Cold Spring Harbor, New York, demonstrated for the first time that in certain kinds of multicolored maize (usually called Indian corn) some of the genes can move from place to place. In a paper published in 1947, she showed that certain elements seemed to be removed from one place and then inserted in others, causing a change in the action of the genes at the new site. At the time, other scientists were puzzled by this report, for the information did not fit with other known facts about genes, and Barbara McClintock's work was mostly ignored. Recent work by other scientists has shown that many diverse organisms, including bacteria, worms, and fruit flies, also have "jumping genes." Finally, the import of Barbara McClintock's early work began to be recognized, and in 1983, at the age of seventy-nine, she was awarded the Nobel Prize.

As early as the beginning of this century, R. A. Emerson of Cornell University had recognized that the varying colors of Indian corn were the result of unstable mutations. Then, in the 1930s, Marcus M. Rhoades, currently at Indiana University, showed that this instability depended on another gene, which he called the *dotted locus.* When the dotted locus was present, the mutation would switch on and off, creating a kernel of corn that was white with spots of purple. Barbara McClintock's contribution was showing that this "switching" gene could move. She identified two factors—one which caused the mutation and one which activated it.

Barbara McClintock at a press conference at Cold Spring Harbor, New York, after winning the 1983 Nobel Prize for her discovery of "jumping genes" and other pioneering work.

Because so many organisms have movable genes, many scientists now believe that these genes may play a major role in long-term genetic change, that is, evolution. In corn and other organisms, these elements which promote mutations are "turned on" when the chromosomes are under stress, and they seem to amplify the impact of the stress, causing increased breakage of the chromosomes. With a large selection of newly created genes, those which are best adapted survive.

RECOMBINANT DNA

One of the most significant breakthroughs in genetic research in the last two decades is the development of techniques that allow scientists to move and replace specific genes on a chromosome. This is sometimes called *gene splicing*, for, if you imagine the chromosome as a piece of rope, it resembles the way in which sections of rope can be spliced together to make one continuous length. Such techniques create the potential to make major changes in cells or even whole organisms at will.

Gene splicing is made possible with chemicals called *restriction enzymes.* These proteins "cut apart" strands of DNA at specific locations. Then other enzymes are used to attach the "loose ends" of the DNA back together. If a new gene which also has "loose ends" is available, it can be inserted into the space where the DNA was cut.

When the DNA has been reformed into its new combination of genes, it is called *recombinant DNA.* When the cell divides, it will reproduce its new form of DNA in the same way that it would have duplicated its original DNA. This technique was developed by two teams of researchers working together—Stanley Cohen and Annie Chang at Stanford University School of Medicine, and Herbert Boyer and Robert Helling at the University of California, San Francisco. When they published their work in a joint paper in 1973, it sparked the beginning of a public debate concerning the wisdom and safety of gene manipulation.

Many people, including the scientists who were working with genes, were concerned that a harmful gene, for instance, one that caused cancer, could be transferred to a cell and then rapidly multiply. If such a cell escaped from the laboratory it could be dangerous. Experiments by Cohen had shown that genes could be transferred not only from one bacteria to another, but from two highly dissimilar organisms. His group had successfully spliced genes from a South African clawed toad, *Xenopus laevis,* into *E. coli* bacteria.

In the summer of 1973, at the Gordon Conference in New England, an annual meeting at which scientists come together to discuss their research, Herbert Boyer gave a lecture on his recombinant DNA experiments. In the discussion that followed, a group of scientists decided to send a letter to the National Academy of Sciences to appoint a committee to study the possible health hazards of recombinant DNA research. When this letter was published in the magazine *Science* that September, the public was awakened to the issues.

After receiving the letter, the National Academy of Sciences in Washington, D.C., decided to consult Paul Berg of Stanford University, who was one of the leaders in the field of recombinant DNA research. He called a meeting of scientists to organize an international conference at which firm guidelines for the use of recombinant DNA could be established. Meanwhile, many researchers decided to postpone their experiments. In February 1975, about one hundred fifty scientists from all over the world met for three days at the Asilomar Conference Center in Pacific Grove, California.

The guidelines of the Asilomar Conference recommended that research on recombinant DNA should be conducted only in special laboratories that were designed so that no harmful products could escape to the outside world. They also recommended the development of special kinds of bacteria that would grow only in laboratory conditions. This would minimize the risk of any bacteria spreading if it did leave the laboratory.

At the Asilomar Conference in 1975, scientists concerned about the risks of gene-splicing research met to establish guidelines for the use of recombinant DNA. Shown here are conference organizers Maxine Singer, Norton Zinder, Sydney Brenner, and Paul Berg, drafting their recommendations.

Recently, fears about the dangers of recombinant DNA have receded. Geneticists have learned that bacteria are unable to translate animal genes they may pick up; that is, although the bacteria may use the gene to make a particular product, the gene itself cannot be reproduced. Also, the *E. coli* which is used for the recombinant DNA experiments today is not able to live outside the laboratory dish.

Today, recombinant DNA techniques provide the means to solve many serious medical problems. By manipulating the genes of the *E. coli* bacteria, scientists can get it to make a specific biochemical product. For instance, drug companies are now using *E. coli* to produce human insulin, a hormone needed by people with the disease diabetes. The human body normally produces its own insulin and uses it to convert glucose to energy. Diabetics cannot produce enough insulin on their own, and without regular injections of it, would die. In the past they have depended on insulin derived from pigs and cattle. However, some people are allergic to animal insulin and the animal supply varies in quality and quantity.

E. coli can also be used to produce human growth hormone (HGH). HGH will benefit the thousands of children who suffer a type of dwarfism caused by the failure of the pituitary gland to make enough HGH. In the past, HGH was so difficult to obtain—it could be derived only from human cadavers—that only 20 percent of the affected children could be treated. Products like human insulin and HGH are just the beginning of the potential use of recombinant DNA to solve many kinds of health problems.

TRAGIC INHERITANCES

When you go to the doctor for a routine checkup, you may be asked to fill out a health questionnaire. On it may be questions both about your own health history and that of your relatives. About 10 percent of all the diseases that affect humankind are known to be hereditary. Once infectious diseases are eliminated, genes may play a role in most, if not all, diseases. Knowing what diseases occur in your family will help you to know if a disease is likely to have been inherited. For instance, if your parents or grandparents suffered from heart disease, than you are more likely to do so as well. If you are aware of a family tendency toward a certain disease, then you are more likely to note its symptoms and to have it diagnosed correctly.

There are over two thousand different genetic disorders known to man. Many of these are caused by a single defective gene. Genetic disease can affect any part of the body. It may be obvious from the moment of birth, or may not appear until late in life, as in the case of Huntington's disease, the debilitating physical and nervous disorder from which singer Woody Guthrie suffered.

Luckily, most genetic diseases are extremely rare, and many strike only certain ethnic or geographic groups. For instance, sickle-cell anemia is found mainly in blacks, Tay-Sachs disease in eastern European Jews, cystic fibrosis in northern European Caucasians, and the thalassemias in Italians, Greeks, and other Mediterranean groups. The following diseases are among those most frequently encountered.

FAMILIAL HYPERCHOLESTEROLEMIA

One of the most common genetic diseases, familial hypercholesterolemia, (FH), affects about one in every five hundred people in the United States. Familial hypercholesterolemia is a hereditary heart ailment in which the victim has an elevated level of blood cholesterol. A child born with FH may have a cholesterol count as high as 1000, nearly five times the normal cholesterol count of 220 in a full-grown man. Although no damage to the heart is apparent at birth, the enormous overload of cholesterol in the blood causes a hardening of the arteries (atherosclerosis), which strains the heart and eventually triggers a heart attack. Children as young as five with two FH genes have had heart attacks.

The FH gene is dominant so that even one gene causes an increase in cholesterol in the blood. People with only one FH gene may not have any difficulty until they are twenty or older, but they often have their first heart attack in their thirties or forties. People born with two FH genes suffer much earlier in life and usually do not live beyond the age of twenty. Luckily, only about one in a million people is born with two FH genes.

Drs. Michael S. Brown and Joseph L. Goldstein at the University of Texas Health Science Center in Dallas have done work that shows exactly how cholesterol is used by the body and for this they were awarded the Nobel Prize for Medicine in 1985. All cells need cholesterol, both for maintenance and, in some cells, to manufacture other products. The cholesterol in the blood is transported into the body cells by special molecules called low-density lipoprotein (LDL) receptors. These molecules combine with the cholesterol so that it can enter the cell. People who have FH genes cannot make the LDL receptor molecules. Therefore, the cholesterol cannot enter the body cells and just accumulates in greater and greater amounts in the bloodstream, where it eventually clogs the blood vessels. For people with two FH genes, the only way to get rid of the extra cholesterol is to clean their blood in a blood-filtering machine. However, people with only one FH gene can reduce their risk of heart disease by altering their diet and amount of daily exercise. Several new drugs also offer hope for relief.

CYSTIC FIBROSIS

Cystic fibrosis is another very common genetic disease and occurs in about one of every twenty-five hundred babies. Symptoms, which include lung infections and an inability to gain weight, begin at birth. People who have cystic fibrosis secrete large amounts of thick mucus, which clogs the lungs and makes it difficult to breathe. Mucus in the intestine prevents food from being absorbed, thus causing the weight loss. Although there is no cure for cystic fibrosis, children with the disease can be helped by a careful diet, medications to aid digestion, and sometimes by a respiratory inhalation machine to aid breathing.

SICKLE-CELL ANEMIA

As early as 1949, scientists Linus Pauling, Harry A. Itano, S. J. Singer, and I. C. Wells found that people with sickle-cell anemia had red blood cells that were different from those in people who did not have the disease. Sickle-cell anemia is an inherited disease that causes red blood cells to change from a flat disk to a crescent, or sicklelike, shape. Normally, the red blood cells absorb oxygen in the lungs and then bring it to tissues elsewhere in the body, where the oxygen is used to convert glucose into energy. When the red blood cells are deformed into the sickle shape, they are less able to absorb oxygen. Sickle cells also break easily and tend to clump together and clog the blood vessels. This can then cause damage to the brain, heart, or kidneys. Victims of sickle-cell anemia frequently die in childhood.

One of the proteins in blood cells is called hemoglobin. It contains iron and gives the blood its red color. It is the hemoglobin that attaches to the molecules of oxygen. Each hemoglobin is made up of about six hundred amino acids. A change in just one of these amino acids changes the cell into the sickle shape. This change, discovered by a scientist named Vernon Ingram, is caused by a minor alteration in one of the genes which directs the formation of hemoglobin.

People who have only one sickle-cell gene do not have severe symptoms of sickle-cell anemia. Although they have some sickle cells in their bodies, the normal gene provides for normal red blood cells which can carry oxygen. However, those who inherit two sickle-cell genes—one from the father and one from the mother—do have severe symptoms. This life-threatening disorder strikes at least one of every 625 black children.

Most people who have the sickle-cell gene live in Central Africa or have ancestors who came from there. This is an area of the world where the often deadly disease malaria is widespread. Malaria is caused by a tiny parasite that lives in red blood cells. Scientists have recently determined that people with one sickle-cell gene have a high resistance to malaria. Thus, by itself, the potentially dangerous sickle-cell gene increases a person's chances of survival, yet, when paired with another sickle-cell gene, it becomes deadly.

An estimated 10 percent of all black Americans have the sickle-cell gene. If two people with the gene marry and have a child, the chances are one in four that the child will get a sickle-cell gene from both parents. There is no cure for sickle-cell anemia. Today, people can have blood tests to determine whether or not they carry the sickle-cell gene.

THALASSEMIA

Another blood disease, thalassemia, sometimes called Cooley's anemia, can occur in a relatively harmless form, like sickle-cell anemia, or can be a severe anemia. The severe anemia, which is often fatal, usually before the age of twenty, causes the body to produce extremely small red blood cells that are very low in hemoglobin. Thalassemia is characterized by onset in early childhood, an enlargement of the spleen, underdevelopment of the body, and a marked anemia.

Red blood cells are manufactured in bone marrow. In a recent controversial study done by researchers at the University of California, Los Angeles, two patients with severe anemia had their bone marrow removed. It was then replaced with bone marrow that had had its defective genes replaced with normal genes. It was hoped that these genes would be able to manufacture normal hemoglobin. However, the results of this experiment were inconclusive.

HEMOPHILIA

Inside the body our blood flows smoothly through our veins and arteries. When we injure ourselves, for instance with a cut or a scratch, then the blood vessels break and the blood flows out of the body. Normally, when blood is exposed to the air it clots or thickens, thus preventing our bodies from losing an excess of it.

In most people, blood clots in five to fifteen minutes. People with hemophilia, called hemophiliacs and, sometimes, "bleeders," have blood that clots extremely slowly, taking from thirty minutes to several hours. For that reason, they are in constant danger of bleeding to death just from minor injuries. They cannot

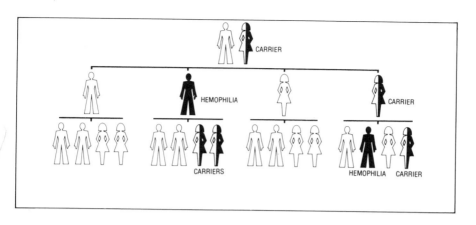

The family tree of a hemophiliac. A peculiar characteristic of hemophilia is that although women are carriers of the gene, only men exhibit the disease.

safely undergo surgery, and even normal dental work can be risky. Many hemophiliacs suffer from crippled joints caused by bleeding and swelling in the joints after they have been bumped or bruised.

England's Queen Victoria had a hemophilia gene. Hemophilia, sometimes called the disease of kings because it was carried by so many members of Europe's royal families, is a hereditary disease. Although it has been known and described since biblical times, the first thorough medical description was made by a Philadelphia physician, John C. Otto, in 1803. He noted that one of its characteristics was that although women transmitted the trait, only men exhibited the disease.

Women carry the hemophilia gene on one of their X chromosomes, but since the gene is recessive they do not have the symptoms of the disease. However, when the defective X chromosome is passed on to a son, there is no matching dominant gene on the Y chromosome to protect him. Thus he gets the disease. A woman who carries a hemophilia gene has a 50 percent chance that her daughters will also be carriers of the disease, and a 50 percent chance that her sons will have the disease. In the past most hemophiliacs died young. Today, they can be treated with blood transfusions.

GALACTOSEMIA

People with galactosemia cannot digest milk because their bodies do not make the enzyme which digests the sugar, called galactose, that is contained in milk. If they drink milk they may suffer from liver damage, mental retardation, convulsions, or even death. Galactosemia is a rare disease caused by two recessive genes. If it is discovered early in life, the person can live a reasonably normal life by changing to a milk-free diet.

NEURAL TUBE DEFECTS

In the development of a human embryo, toward the end of the first month, a long, flat structure along the back of the embryo turns up like a sheet of paper being made into a roll. The enclosed tube is the beginning of the backbone, spinal cord, and brain. Sometimes, due to a genetic disorder called spina bifida, the roll does not completely close and parts of the spinal cord and nerves are exposed. Although babies born with spina bifida can have the opening closed with surgery, they may become paralyzed. Some such babies also suffer from hydrocephaly, or water on the brain, and many are mentally retarded. Only one to two babies out of a thousand are likely to have neural tube defects. Through a special series of tests, these babies can be identified while still in the womb.

GENES AND CANCER

One of the most dreaded diseases of modern times is cancer. Although extensive research has greatly increased our understanding of cancer, it is still unknown what exactly triggers the uncontrolled growth of cancerous tumors. In some families, where certain kinds of cancer occur frequently, it seems that there may be a specific cancer gene which is passed from one generation to the next.

The first indication that there were cancer genes was shown in 1910 by Peyton Rous of the Rockefeller Institute for Medical Research, when he found that viruses from chicken tumors could cause the same tumors when injected into other chickens. Viruses, which consist of relatively short strands of DNA, invade cells and use the cell's resources to multiply their own genetic material. In 1966, at the age of eighty-five, Rous received the Nobel Prize for his work.

Recently it has been shown that the genes that produce cancer are found not only in viruses and cancer cells, but in normal cells as well. It may be that cancer genes are among the essential parts of the cell's genetic material, directing normal cellular activities under normal conditions, but betraying the cell when disturbed by carcinogens (agents in the environment which promote cancer).

Cancer genes, also called "oncogenes," seem to be activated when their original chromosome breaks and then later recombines with the wrong part of another broken chromosome. Jorge J. Yunis of the University of Minnesota, in a report made in November 1984, suggested that when the cell's chromosomes break at certain identifiable points, the development of cancer is more likely. He also suggested that the potential for breakage may be influenced by diet, although this aspect of his report is very controversial.

One of the problems of treating cancer is that cancer genes seem to be no different from those found in noncancerous cells, so that any method that destroys them, destroys healthy cells as well. However, there are some new products derived from biotechnology that may be able to kill cancer cells without harming normal cells. In 1975, Cesar Milstein and Georges Kohler at the Medical Research Council laboratory in Cambridge, England, created cells called monoclonal antibodies, which are able to repel cancer cells. Normally, when the body is infected with a foreign substance, the immune system produces antibodies to fight the invader. When cancer cells multiply quickly the body cannot provide enough antibodies to fight them. Milstein and Kohler fused antibody-producing cells and tumor cells, thus making cells, called *hybridomas*, that will reproduce continually and create a steady flow of antibodies. Other scientists have attached radioactive molecules to monoclonal antibodies. When these antibodies bind to cancer cells, the radioactivity kills the tumor. Radioactive antibodies have also begun to be used as cancer detectors. The problem of cancer has not yet been solved, but as scientists increase their understanding of how the cell works, both in normal and abnormal states, we are getting closer to learning how to cure this dreaded disease.

GENETIC COUNSELING

Most genetic diseases are rare. However, a couple may have a relative or even a child of their own with a genetic disease. They may worry that if they have children or more children, those may have the disease, too. Today, such people can consult a genetic counselor. Usually the counselor is part of a team that includes a medical geneticist, a cytologist (a specialist in the study of cells), a clinical pathologist (a specialist in the study of disease), and a social worker. These experts can use the most modern scientific techniques to try to determine which genes the mother and father carry. If the woman is already pregnant, they may be able to determine which, if any, harmful genes may have been passed on to the child.

A person receiving genetic counseling may first be asked to provide a detailed family medical history. If a genetic disease recurs within the family, it may be inherited. Potential parents may then have blood samples taken from their arms. The blood is placed in a laboratory dish on a substance that stimulates the growth of white blood cells. Three days later, the cells are removed, placed on microscope slides, and the chromosomes are stained to make them more visible. The chromosomes are then photographed so that they can later be sorted into pairs, a process called "karyotyping." Since people have forty-six chromosomes, they are sorted into twenty-three pairs, the twenty-third pair being the sex chromosomes, either XX (female) or XY (male). The chromosomes are then examined carefully for abnormalities. Sometimes the blood is studied for disorders of specific single genes, as in Tay-Sachs disease, sickle cell anemia, thalassemia, galactosemia, or hypercholesterolemia. Various biochemical analyses, rather than chromosomal analysis are used in such cases.

In order to examine the chromosomes of an unborn child, a doctor must use a procedure called *amniocentesis*, in which a long, hollow needle is inserted into the mother's uterus so that a small amount of fluid can be extracted. This fluid, which surrounds the fetus inside the uterus, contains cells from the growing fetus: Like the blood cells from the parents, these fetal cells can be grown in the laboratory so that the chromosomes can be examined. Such tests can also determine the sex of the child, a feature which is particularly useful when looking for genes associated with sex-linked diseases.

Amniocentesis is usually done in the sixteenth week of pregnancy. A new technique called chorionic villi sampling, which can be done between the eighth and twelfth weeks, may eventually replace amniocentesis. In the chorionic villi sampling procedure a small portion of the membrane that surrounds the fetus is extracted. Because the sample can be taken earlier in the pregnancy and because it can be analyzed more quickly, it allows people to make an earlier decision.

Sometimes, when the chromosomes are examined, an extra one is counted. Down's syndrome, also called *mongoloidism* is one of the disorders associated with having an extra chromosome. People with Down's syndrome generally have a skinfold at the corner of the eye, which gives them a characteristic mongoloid appearance; they also are mentally retarded to some degree and may have other health problems, including congenital heart disease. Some types of Down's syndrome are passed from one generation to the next and run in families; others result from an extra chromosome that is present in the egg or sperm cell. The chance of this increases with a woman's and, to a lesser extent, a man's age. Today most doctors recommend that women over the age of thirty-five who become pregnant undergo amniocentesis.

Many ethical questions surround genetic counseling. Counselors can help people understand the risk of bearing a child with genetic disease and how that might have an impact on their lives, but they cannot make the decision about whether or not the parents should bear that child.

Recently, techniques for widespread screening of people to determine whether they carry a disease gene have been developed. These tests use radioactive tracers and biochemical techniques to detect carriers of Tay-Sachs disease, Neimann-Pick disease (for which there is no effective treatment and from which most children die before the age of five), and at least thirty other disorders. (Although carrier detection is possible for Neimann-Pick disease, widespread screening is not possible.) In diseases that require a pair of genes to be expressed—one from the mother and one from the father—people who carry one harmful gene would otherwise never know that they had such a gene. Yet, if such a person should have a child with another person who carried the same gene, then they would have a one-in-four chance that the child would have the disease. With guidance from genetic counselors, information from genetic tests can help people make choices about having children.

CHAPTER

GENES AND AGRICULTURE

A practical knowledge of the hereditary process came long before its mechanism was understood. Archeologists have discovered that as long as seven thousand years ago farmers in Central America were improving crops of corn by planting seeds of corn plants that had developed preferred characteristics. Over six thousand years ago, the Chinese learned how to develop superior strains of rice. An ancient Babylonian tablet shows a pedigree of a family of horses through five generations, with detailed information about height, length of mane, and other traits, revealing that they had some knowledge about how these traits were transmitted.

Farmers and gardeners have continued to practice this type of selective breeding in both plants and animals. Each time an individual plant or animal appeared with a desired characteristic, it was bred again to produce more with similar traits. For instance, at harvest time, farmers would select heads of wheat that had the most or largest kernels and save them to use as seed for the following year.

Animal breeders have also been able to use a knowledge of genetics to improve their stock. Today breeders can produce sheep with better qualities of wool, pigs with more or less fat, as desired, and cattle with a higher quality of meat. Other new techniques such as artificial insemination allow animal breeders to greatly increase the offspring from genetically superior animals.

GEORGE SHULL
AND HYBRID CORN

After 1900, as Mendel's laws began to be understood, crossbreeding became much more carefully controlled than it had been in the past. A leader in the field of plant hybridization was a geneticist named George Harrison Shull, who worked at the Station for Experimental Evolution in Cold Spring Harbor, Long Island. Shull joined the Station in 1904, and his first plots of corn were planted there in order to develop for visitors a display that would illustrate Mendel's laws of inheritance, which had recently been rediscovered.

As Shull began his work he was aware of others who had experimented with plant breeding before him. One of these was Charles Darwin, who had observed that crossings of plants from unrelated strains tended to produce stronger hybrids than those from related strains. The first person to do controlled studies of crosses for various strains of corn plants was a follower of Darwin, William James Beal. In 1879, at the Michigan Agricultural College, he planted two strains of corn, both in the same field.

Beal conducted his experiment in the following way. First he removed the tassels from all the corn plants of the first variety. The tassels of corn are the male part of the plant, and they produce the pollen which fertilizes the ear shoots. Normally, corn plants can be self-fertilizing. However, by removing all the tassels from one variety, Beal prevented those plants from self-fertilization. Instead, those ears became fertilized by the pollen of the other variety, which was blown by the wind from the tassles of the corn planted in adjacent rows. Then, after the ears had matured into full kernels, the ears from the first, detasseled, plants were harvested. These kernels were the seeds for the

George Shull and his plot of corn at Cold Spring Harbor, New York, circa 1905. Shull's work provided a biological basis for breeding hybrid corn.

Diagram illustrating capacity for growth in corn. The figures at the top represent the average height in inches for the different generations identified at the bottom.

hybrid corn to be planted the following year. This process of detasseling is still used today to produce hybrid corn varieties.

When Beal planted his hybrid seeds he found, as expected, that the plants had higher yields. Beal did not attempt to purify the strains or to determine the genetic basis of his hybrids.

George Shull and another scientist working at the same time, E. M. East at the Connecticut Agricultural Experiment Station in New Haven, were the first to isolate pure strains of corn. Over several years Shull bred the progeny of a single ear of white corn. Then he crossed two pure lines to produce a hybrid. He found that when these seeds were planted they produced highly uniform and higher-yielding plants than either of the two pure strains.

Shull developed the concept of hybrid vigor—the idea that increased yield, vigor, and rate of growth of plants comes from the mating of unrelated parents. Ordinarily, when plants or animals are bred again and again with others of a similar genetic makeup—a process called *inbreeding*—the resulting offspring become increasingly uniform, accenting the desired characteristics. As the line becomes more and more inbred, the weaknesses appear more often as well (see illustration on page 60.). In pure strains of corn, for instance, plants in successive generations become smaller and weaker. By crossing the corn with an unrelated strain—a process called *outbreeding*—the result is a stronger, healthier, larger plant.

George Shull's work provided, for the first time, a biological basis for breeding hybrid corn. However, although the first generation of such crosses was vigorous, he found that the second was not, and thus seed manufacturers did not find this a practical method to produce stock. Then, in 1918, another geneticist, Donald F. Jones at the Connecticut Agricultural Experiment Station, tried crossing two pure hybrids. This method—called the *double cross* because it crossed four pure strains—resulted in seeds that were vigorous and that produced a 20 percent better yield at harvest time. Finally, a practical source for seed corn had been established. The double-cross method was used for many years to produce seed corn. However, today, the development of better inbred lines has made single crosses better than the best double crosses.

The use of hybrid corn seed has had a dramatic effect on corn production both in the United States and around the world. In the United States, between 1930 and 1979, the yields of corn per acre grew from 21.9 to 95.1 bushels per acre. Although only two-thirds of the land that was used in 1930 for corn is now in production, it produces twice as much corn!

The methods developed by Shull to increase corn yields have been used with other food crops such as sorghum and rice. As the population of the world increases, such improvements are essential to keep the world fed.

NEW METHODS OF PLANT IMPROVEMENT

Many people feel that improvement of crop production through hybridization has reached its limits. Maximizing yields requires fertilizers which have become increasingly expensive. However, several new methods of genetic manipulation may be able to produce highly improved plants that use less fertilizer, or even make altogether new plants.

One method that has been successful in producing dwarf varieties of apple, pear, and other fruit trees is cloning. Scientists take a few cells from adult trees and put them in a growth solution. There the cells begin to grow into new plants genetically identical to their "parent." Such dwarf trees are often preferred by the commercial grower, for they bear fruit sooner and are easier to harvest.

Sometimes, when identical cells are grown in laboratory cultures, a few of them develop new traits. This phenomenon is called *somaclonal variation*. Scientists have taken advantage of somaclonal variation to produce new varieties of tomatoes,

In a somaclonal variation experiment, these little tomato plants were regenerated from cells in tissue culture.

including the thick-skinned, easily harvested, but rather tasteless variety found in most supermarkets.

An experimental technique which may produce new plants in the future is called *protoplast fusion.* Here, the walls of two different kinds of cells are broken and the two cells are joined to become one larger cell. Such a method has been used to join a tomato cell and a potato cell. The resulting product is a potato that smells like a tomato! Fusion works only a small percentage of the time, but it is possible that this method could be used in the future to produce disease-resistant plants.

One of the most appealing, but difficult, new techniques is gene splicing, in which a piece of DNA is inserted into the plant cell, providing it with new genes. Theoretically, such a method would allow plant breeders to insert into a crop line any desirable gene. For instance, if a gene for drought resistance or salt resistance could be inserted into a food crop plant such as wheat or barley, it could open up areas for agriculture where plants normally cannot grow.

Some of the most serious problems facing mankind are hunger, sickness, and energy shortages. All of these have potential solutions through genetic engineering. One of the problems in raising basic food crops such as rice and wheat is the rising cost of fertilizer. Some plants, such as those in the bean family, are able to produce their own nitrogen fertilizer with the help of bacteria that live among their roots. If scientists could insert a nitrogen-fixing gene into a wheat or rice plant, this would greatly reduce the cost of growing food. Although this has not yet been done, scientists are currently working on several solutions to these problems.

CHAPTER

7

THE FUTURE

A recent newspaper headline read, "Gene Therapy for Some Diseases Called Close at Hand." The article which followed described the establishment of a multimillion-dollar Center for Molecular and Genetic Medicine at Stanford University. The center, which will open in 1988 and be directed by Nobel Prize winner Paul Berg, will attempt to apply the knowledge gained from genetic research to clinical problems.

Of the two thousand single-gene diseases that have now been identified, replacement genes for approximately one hundred have been cloned. Although many of these diseases are rare, they account for a considerable number of childhood admissions to hospitals. These diseases result when the gene that controls the manufacture of a specific protein is lacking, or produces the protein in too large or too small amounts or produces an ineffectual protein. If the correct gene could replace the defective one, then the condition could be cured.

One technique that will be used at the new center involves using a special kind of virus called a *retrovirus* to introduce a healthy gene into the cells of a patient whose cells do not have that gene. One experiment of this type, planned by Dr. David Martin at the University of California, San Francisco, will replace a gene needed for the normal function of T cells in the immune system. The new gene will be placed in bone marrow, which will

then be transplanted into the patient. The body needs T cells to fight off infections. During the near future, several other medical centers in the United States will also be trying new solutions to single-gene diseases.

THE GENE BUSINESS

Before 1981, the creation of a gene segment in the laboratory took a team of scientists and technicians four to eight months of painstaking labor. Today, a "gene machine" can do the same job in a day. A gene machine is an automated genetic engineering laboratory which, at the push of a few buttons, can synthesize a fully workable gene segment. The segments can then either be spliced together or inserted into the DNA of a living organism. The gene machine allows scientists to synthesize a natural gene, to modify a gene, or even to invent a completely new gene.

To use a gene machine a person types onto a keyboard the genetic code, that is, the amino acid sequence of a protein. Then a microprocessor stores the information. The microprocessor controls a maze of tubes and valves that emit appropriate amounts of the gene's subunits, nucleotides, into a reaction vessel. There the DNA chain is gradually built like so many beads on a string.

New techniques such as gene splicing and new tools such as the gene machine have helped propel genetic engineering into a multimillion-dollar business. In the last decade over two hundred companies in the United States alone have entered the field,

A state-of-the-art DNA synthesizer, capable of performing three simultaneous DNA syntheses. The instrument is controlled by menu-driven software accessed by a touch-screen monitor.

making products that range from a vaccine against a cattle disease called *scours*, which normally kills about 10 percent of United States cattle a year, to a protein that can prevent heart attacks by dissolving blood clots that obstruct coronary arteries.

Because of the potential for huge profits, the competition between genetic engineering companies is fierce, and this has led to secrecy. Such companies have also drawn researchers away from universities and medical centers, where, although the research tends to be more theoretical, the time, knowledge, and even equipment and specimens are openly shared. The conflict between scientific freedom and corporate profits is a problem that will have to be resolved.

The field of genetic research has expanded so fast in the last few decades that it has been more like an explosion than a linear development. From the discovery of the structure of DNA in 1953 to the ability of present-day scientists to reconstruct DNA to make specific lifesaving biochemical products, a great deal about the function and manufacture of genes has been revealed. Yet, there is still much to learn, and it is hard to believe that this scientific revolution began a little more than a hundred years ago.

Who would have thought that a substitute high school science teacher who was never able to pass the examination for his full teaching credential would have laid the foundation for modern biological science? Johann Gregor Mendel was that failed science teacher who experimented with peas in the monastery garden in Brunn, Austria, providing the first systematic evidence that physical traits are passed on from one generation to the next—that is, that children inherit their physical characteristics from their parents.

Today genetic research holds the answers to many problems facing mankind, including increasing the world food supply through agricultural research and reducing birth defects through genetic counseling. Many previously mysterious areas of science can now be understood through the study of genetics. The legacy of Mendel's pioneering work will reach far into the future.

FURTHER READING

Bornstein, Jerry and Sandy. *What Is Genetics?* New York: Julian Messner, 1979.

_____ *Discovering Genetics.* New York: Stonehenge Press, in association with the American Museum of Natural History, 1982.

Dunbar, Robert. *Heredity.* New York: Franklin Watts, 1978.

Facklam, Margery and Howard. *From Cell to Clone: the Story of Genetic Engineering.* New York: Harcourt Brace Jovanovich, 1979.

Hyde, Margaret O., and Lawrence E. Hyde. *Cloning and the New Genetics.* Hillside, N.J.: Enslow, 1984.

Klein, Aaron E. *Threads of Life: Genetics from Aristotle to DNA.* Garden City, N.Y.: Natural History Press/Doubleday, 1970.

Lampton, Christoper. *DNA and the Creation of New Life.* New York: Arco, 1983.

Patent, Dorothy Hinshaw. *Evolution Goes On Every Day.* New York: Holiday House, 1977.

INDEX